Bottom

By

Ronald L. Carbery

Introduction

In this document I try to present a variety of specific philosophy summaries that make up what our country and society is or should be comprised of. In it I include what it means to be a patriotic American, what honesty and integrity should mean to those people who should possess them, what it means to be a team player and to cover the backs of those people who are working to make this a better America and a better world. I also discuss what the results were from being involved with subprime mortgages as controlled and handled by Fannie Mae and Freddie Mac. I also have a discussion of what usury is and how it used to be handled at the turn-of-the-century in 1900 compared to how people are charged such ridiculously high interest rates now and why they should not be. I also give a discussion of the biosphere and how were doing a real job on our own environment and how long we can expect before we start paying the consequences for our excessive use of our resources.

Being an "American"

On being an American, probably the first way to approach this is to consider why and how this country of ours was formed to begin with. We need to do a little reviewing of our initial history of our country and reflect on why those people living here in the middle 1700s decided that they needed to disassociate themselves from their king who resided in Great Britain Way across the ocean. There were several reasons why they decided that they want to remove themselves from the British Empire. Probably the first one on their mind or the first reason was that they were being taxed heavily by their foreign king and they had nothing to say about it. In other words it was taxation without representation and this made them quite angry. They were working hard trying to make ends meet so they could feed their families, cloth their families and keep their dwellings in livable condition while at the same time being taxed heavily for everything they did. The taxation became such a burden that I got to a point where they couldn't stand it anymore. So the colonists got together and decided they needed to take action and the action they decide to take was to secede from England. And, of course, that's where the revolution came in. That's when the colonists decided to elect George Washington as their first leader and they decided that they would fight the British in order to remove themselves from the control of the King.

Since the time of the revolution the United States of America has fought in many wars and lost many lives as a result. The war of 1812, the Spanish-American war, the first world war, then the second world war, then the Korean War, then the Vietnam War, then Desert Storm, then the Iraq war, and then the Afghanistan war. This last war of course, was brought upon us as result of the attacks on us on September 11, 2001 by the Muslim extremist. We all remember that in these wars and especially the attack on

9/11. We all remember where we were and what we were doing when we were told and then observed the twin towers in New York City being struck by aircraft and the Pentagon being struck by an aircraft and the aircraft crashing in Pennsylvania due to the efforts of some onboard brave souls.

Who could ever forget the horrible scenes of those two towers eventually collapsing on themselves and all of those poor people being killed as a result of this? I remember what I was doing when this happened, I was a contractor for DOE working in the Forestall building in Washington DC. I looked out the window across the Potomac and I could see the flames billowing out of the Pentagon from that window. All the people in the office were incredulous as to what had happened. Within about a half an hour all the people working there were told to try to make their way home however they could and eventually we contractors also left. I can remember walking from the Forestall building to the Memorial Bridge and crossing the Potomac and then walking on home until I got close enough to my own home where I hitched a ride to where I lived in Maryland. I'm sure most other people had similar experiences that they can remember.

This was one of those days nobody forgets.

If you have lived here all your life and been told of the history of our country and have reviewed its trials and tribulations and seeing how the people in our country the government has try to treat all individuals with respect, then you get some understanding of the way a patriotic American should think and also act in this country. It always helps to remember that our

Constitution starts out with quote we the people unquote. It pays to remember that it is the people who give the power to the government not the other way around, it's also important to remember that the government does not make any money it is the people who make the money and it is the government who taxes the people to obtain the funds it needs to operate as a government for the people. This must be remembered because it's the key to why the government must try to control its spending so as not to overburden its people with overburdening taxes. The people must be allowed to keep as much of their hard-earned money as possible in order to continue to improve their lives to expand their businesses and to increase the tax base for our country.

It would certainly seem that the current administration either has no concept of these ideas or has decided to ignore them because of its own specific agenda.

Recently, the "Tea Party.net" has produced a petition for which it is trying to gain signatures to impeach President Obama and remove him from office. In that discussion they mentioned that America is under attack from within. And they say over the last 4 ½ years that our nation has been transformed and for the worse so much so that one can hardly recognize it. They say we have a corrupt Chicago politician in the White House who's bleeding our nation to death and that since he won the reelection with the help of the uninformed electorate and a totally compliant news media many details of further corruption have been revealed. They are saying for example that he is purposely lied about the Benghazi atrocity and instructed others to lie also and intimidated any whistleblowers who were guilty of nothing more than a desire for justice and the truth. They say that while our Libyan ambassador was burned and others were slaughtered he went to bed not even caring enough to make one phone call as to the status of their seven hours of hell. And as we all know from the news we have heard on Fox prior to these attacks ambassador Christopher Stephens had made multiple requests for extra security Benghazi and these requests were refused by the State Department. And even despite the fact that the British Embassy in Libya had been previously attacked only months earlier by Islamic terrorists. Obama has use taxpayer funds to run TV ads in Pakistan pushing and promoting the lie that the Benghazi tax were caused by a YouTube video which offended the Muslims. Now as of May 2013, we find out that the military knows exactly who the perpetrators of this attack were and has been keeping them under surveillance since last October in 2012. And we find out that the reason that they have not been arrested as of yet is because Pres. Obama does not want to put them into Guantánamo as terrorists because he promised his constituency that he wouldn't add any more people to the Guantánamo facility. But the United States does not have enough evidence on them to bring them to the United States for trial and so therefore they been left free to roam about in their own countries simply because Obama will not have them arrested and put in one Guantánamo and in detention. This is absolutely outrageous of course.

However, Pres. Obama continues to get away with his ongoing unlawful activities and nobody in Congress seems to have the guts or will to do anything about it.

As for the Sequester cuts, which has since been shown to be an insignificant cut in the rate of growth of government spending, Obama close the White House to tours for our children while at same time sending hundreds of millions of taxpayer funds to known terrorist groups the Muslim brotherhood in Egypt also giving them F-16s and tanks.

It would seem he treats our foreign friends like enemies and rewards our enemies as if they were our friends.

In addition, we now have the IRS scandal wearing conservative organizations such as the tea party were targeted by the IRS for audits and harassment when they tried to apply for tax-exempt status. They were asked all kinds of questions even to the extent of who they talked to who their friends were if they had ever had any speaking engagements and what they discussed at the speaking engagements. It's a simply unbelievable overuse of federal power to intimidate people. Anyone who spoke against the Obama administration and any of their activities were targeted apparently. None of this harassment was directed at liberals or progressives of course because naturally they're all Obama's friends and supporters.

We also have the AP debacle where the Justice Department has been eavesdropping on all the telephone calls and emails from the AP reporters to any sources whatsoever even including the United States Congress. This is another entirely big abuse of power by the Obama administration. They were trying to find out who was on their side and who was not who was against them. And it goes on and on.

In addition, he has bridled our nation with over 20,000 pages of additional and incoherent rules regulations and mandates now required as the result of implementing his new "health mandate" in the form of Obama care which gives the federal government the power to force United States citizens to buy something that only the government approves of.

The tea party goes on to say Obama has divided the country like no other time in history since the Civil War that he has pitted Men against Women and gay people against straight people and used un-American Marxist "Class Warfare" that has no place in a free America. They go on to say he has cause racial strife by unjustifiably labeling anyone who disagrees with his USSR style governance as a racist. In other words, if you have conservatives ideas and think the government should tax less, spend less and be smaller in scope and size, then you must be a racists.

The tea party goes on to say that Obama has degraded our people at every opportunity with his apologizing for America on foreign soil. His bowing to the Saudi Prince humiliating our nation then to the Mexican people where he blames the American people for his illegal gunrunning operation "fast and furious". He apologized to the Mexican people for US sovereignty while inferring that the lower region of our country still belonged to them. These actions are those of a "Traitor" to this great nation. He continually and on numerous occasions abuses the power of his office. He tried to cover up his minions in his own administration regarding crimes in the fast and furious gunrunning operation. It was the Obama administration which gave guns to Mexican drug gangs in an attempt to later broach and attack the Second Amendment and when questioned about this activity which was apparently initiated by his attorney general and which he obviously knew all about, he then proceeded to plead calling it "executive privilege".

He keeps adding to the already unbelievably huge national debt. He increased it by about 60% during his first term over $6 trillion while still insisting that even then we must spend even more. He has failed to get any of his budgets passed even during the period . When his own party held both houses of diarist. At that time his budget was defeated for 14 to 0 in the house and 90 920 in the Senate he didn't even received one vote from his own party.

Then he wasted 862 billion on a stimulus plan which was supposed to create shovel ready jobs and then later on he admitted there was no such thing as a shovel ready job available at that time.

The tea party states Obama has promoted voter fraud at every possible occasion and failed to convict proven voter intimidation thugs based upon the skin color of the lawbreakers.

They say he has advertised and promoted American food stamp programs over Mexican airwaves to Mexican citizens an attempt to swarm our country with millions more in illegal aliens and further bankrupting our country with these tactics.

He has sued several states repeatedly trying to stop them from implementing the laws of their own citizens had approved violating the 10th amendment in the process while completely obliterating the constitutional principle of state sovereignty.

He bypasses Congress every chance he gets by using the federal bureaucracy to inflict massive and crushing laws on the people even though it's totally unconstitutional to do so.

He has also pointed dozens of unconstitutional czars, and these people are unchecked by any balance of powers but they serve at his pleasure and mainly to harass the rest of us.

From what we've learned Obama has created an enemies list, just like Richard Nixon did, which includes any individuals, companies, or industry that disagrees with his destructive policies. He tries to control the Free Press and tries to tell Americans which media sources that they should listen to.

The tea party points out that Obama embarrassed and harmed our relationship with our ally Great Britain by sending an official US delegation to socialist dictator Hugo Shabazz's funeral but by not sending anyone to the great defender of liberty the great former British Prime Minister Margaret Thatcher's funeral.

They also point out that he uses every tragedy which comes along to attack our constitutional freedoms as, for example, using the Sandy Hook's elementary school shootings to infringe upon our nonnegotiable Second Amendment rights and of course were all aware of his use of the Benghazi terrorist attacks and blaming them on the First Amendment in which a YouTube video critical of Islam was placed on YouTube, all the while knowing full well that this was an orchestrated Islamic jihadist attack. He even had his Secretary of State Hillary Clinton pushing the same story. Then he had his ambassador to the UN, Susan Gates, go on five different news stations and give the same story about the nasty Islamic video. His apparent unconcerned attitude about his own Amb. Chris Stevens, and the other three patriotic Americans who were killed didn't even keep him up that night because he went to bed during the whole situation. It makes one wonder if he's concerned about his citizens at all?

And he has all he has unconstitutionally attacked our sacred freedom of religion a God-given right guaranteed by our founders in the First Amendment.

The tea party thinks Pres. Obama is the most corrupt president in the history of the US and that his actions are against everything this country was founded upon and stands for. He is a danger to America. We are therefore well within our rights to call for his impeachment and removal from office and I totally agree with them. I think it would be in the best interest for the good of the nation and to keep the legacy of our founding fathers alive to remove Obama from office.

Honesty & Integrity.

With respect honesty and integrity, I look back on my career and especially my career in the Air Force. When you first going to service you have to go through what they called basic training. This basic training normally takes about three months and in this three months they try to teach all of the enlistees to become team members and respect each other and to do their jobs as they're told to do them. They also give you physical training to try to get you into some kind of physical shape but along with this they give you some book training by teaching you some of the history of the United States including the history of the revolution and the Constitution of the United States and the basic laws which are included in these documents.

I think that most people would agree with me that many of our countrymen have no idea of the history of this great nation, they don't seem to be learning about it in the schools they go to nor do they seem to learn about it from their own parents or peers. It was seen that the younger generation as it has no idea of what happened at Pearl Harbor in 1941 or knows much about Nazi Germany or the Holocaust and the extermination of over 8 million Jewish folks. They seem very knowledgeable about iPods and iPads and play stations and cell phones and Facebook and twitter, etc. But they seem to have lost touch with their own country and its history and quite often with their own people. When I was in the Air Force I decided to go to officer candidate school, usually referred to as OCS. This was a six-month course where for the first three months you were lower-class and if you past that the last three months you were upper-class. If you manage to pass the upper-class you were commissioned a second lieutenant in the United States Air Force. During the six-month course you learned that you are word was your bond and that anything you said you would be accountable for. In other words anything you say you have to say what you mean. You also learned that if you signed something it was totally legal and you would be also accountable for that as well you were also taught that if you were told to do something or to be someplace at a specific time you had better be there or do what you're told because if you didn't you would be kicked out of the program very quickly. I know of one case where one of our best candidates came back to base 10 minutes after midnight one night and he lied about it and because of this he was kicked out of the program and this was after 5 ½ months of training.

I submit that Obama could not have survived this OCS training because he lies and he cheats and he apparently has no respect for other people at all.

In line with the fact that Obama has no empathy or understanding of real people and how they should be treated reminds me of when I was in the United States Air Force and was a Detachment Commander in Germany. I periodically promoted Airmen and Women to supervisory positions. I wrote a "Supervisors Handbook" to give them some training and a source to review if they had questions about their jobs. I'll include it here just because it makes good reading and sense in the context of trying to work with people, especially really good people.

For most jobs out there the requirements of the supervisor are very seriously needed and so of course people are not just born to be supervisors, they must be trained to be supervisors, They must have the proper schooling and education in order to develop the proper characteristics to be a good supervisor. As a result there needs to be some kind of a training course given to these people when they first come on board if they're going to be a good supervisor.

"Supervisors Handbook" prepared by "Communications Officer" Detachment 200

INTRODUCTION

1. You have been selected as a supervisor because you have certain personal and military qualifications.

a. You are willing and capable of assuming responsibility.

b. You have demonstrated ability to lead and to get along with men and women.

c. Your conduct both on and off duty sets a good example.

d. You are qualified in your Air Force specialty.

2. As a supervisor you are responsible for the men and women of your section. You must make sure that they perform the duties required of them. You are responsible to see to it that they are trained to do the job assigned. You must make sure that they are fed. This means that their duties must be so arranged that they can go to the dining Hall and eat. You must see that they were have proper clothing; that they keep themselves clean and have

haircuts and proper grooming. It will be your responsibility to see to it that the sections property and equipment is properly maintained. You must supervise their work so that you know that they perform the job safely and well.

This may seem like a large order, but, every leader or supervisor has these responsibilities. In addition, remember that your men and women must bring their problems to you, whether these problems are personal or in connection with the assigned job. If you cannot solve the problem, then refer it to your supervisor.

3. On the pages of this handbook are listed your major responsibilities as a supervisor. Specific responsibilities and tasks will be outlined in your position description. To be a successful supervisor you must carry it them out and well. You will be watched closely and your success will be measured. The success of your mission is the summation of the successes of all the sections in the organization.

It is suggested that you study the airmen's handbook, and AFM 35 – 15, Air Force leadership. A knowledge of these publications will aid you in performing your duties

4. My wish is for your success as a supervisor.

OIC COMMUNICATIONS

"SECTION I

REQUIREMENTS OF LEADERSHIP

1. This section outlines briefly those things which you as a supervisor should know about yourself, your men and women, and your job.
A diligent study of these requirements will give you a good understanding of what is required of a supervisor.

2. Know Your Self:
a. Be sure of your responsibilities to your commander.
b. Learn and practice good personnel management.
c, Know and practice the principles of leadership!
d, There is no loyalty less than complete loyalty to your superiors as well as to your subordinates.
e. Don't be a "yes man" - but worse yet - don't be a "chronic griper".
f. Study yourself and your faults, and strive constantly to correct them.
g. Organize your thoughts as well as your acts. Don't act on the spur of the moment. THINK!

3. Know Your Men and Women:

a. Know each one by name.
b. Be sure you can account for each one at all times while at training, on detail, on the job, or during off duty hours.

c. Keep track of your men and women. If he or she is absent, know why he or she is absent. In addition, let your subordinate know where you are, during your absence.

d, Whenever possible, enjoy and maintain the confidence of each individual in your section; encourage him or her to discuss their difficulties with you - help them.

e. NEVER betray the confidence of one of your people.

f. Carefully review the records of each of your people immediately after he or she is assigned to you. Assure yourself that their records correctly reflect their duty assignment and capabilities.

g. Watch each man or woman carefully and determine their human failings or deficiencies. This will help you guide him or her and will eliminate AWOL's, Disciplinary Reports, and injuries to members of your section while at work.

4. <u>Do You Know:</u>

a. The principles of promotion of airmen?

b. The principles and procedures for the reduction of airmen?

c. The requirements for each Air Force Specialty (AFS) authorized and used in your section? (See AFR's in 35 series and the Unit Manning Document.) Does each member of your section know his Air Force Specialty Code (AFSC)?

d. How to maintain training records and to accomplish appropriate up-grading of your men and women.

e. All up-to-date instructions and Standard Operating Procedures regarding operations of your section and squadron?

f. All scheduled inspections and what will be inspected? (Check section bulletin boards.)

g. The capabilities, limitations, and mission requirements of your section and its equipment?

h. The function of your section in relation to the mission of the Air Support Operations Center.

i. The function of your section as it pertains to:

(1) Detachment 200?

(2) The Squadron?

j. Your exact responsibilities?

k. To whom you report?

l. Each days work or training schedule?

m. What is going on in the world? Are you up-to-date?

5. You <u>Should </u>Read:

a. All section publications.

b. A good book on leadership or personnel management.

c. All squadron publications.

d. Ground safety directives.

e. A daily newspaper and a service publication like "Air Force Times", etc.

SECTION II

PRINCIPLES OF GOOD SUPERVISION

1. This section will explain to you some of the principles employed by good supervisors. As a supervisor you not only have to lead your men and women, but you must supervise their work, training and other activities.

2. Teaching & Briefing. Teaching is the art of imparting knowledge of skill in some subject to another person or group of persons.

How to do some job or make something is a good example.
Briefing
is giving a short accurate summary of what is to be done or accomplished. You as a team leader must learn and practice the art of teaching and briefing your men and women in the proper manner of doing a job. It is necessary that your instructions be clear and concise.

3. Checking the Job., Just because you have instructed one of your men or woman on how to do a job, and then requested or directed him or her to do it, do not sit back and assume that the job will be done.

You must check and make certain the job is done. Also just because you have instructed one of your people to do a job and briefed them how to do it, do not assume that doing the job is no longer your responsibility. You will be the one held accountable if the job is not done properly, not the man or woman you instructed to do the job. You are paid to supervise and to take responsibility for an assigned job.

4. Inspections. You must make frequent and thorough inspections to determine that your section is functioning properly; that your people are living cleanly, that your section is neat and is being maintained in a military manner. Remember that you must insist that your men shave daily, the women keep clean and dress cleanly and keep their bodies and living quarters clean. Efficiency is directly related to cleanliness.

"Spit and Polish" never hurts anyone.

5. Conduct. Your conduct is directly related to the respect you receive from your men and women. If you do not have the respect of your people, you will not succeed as a supervisor. For this reason you should not gamble or drink with them. Swearing accomplishes nothing and loses respect. Above all, you must set the example in cleanliness and conduct.

6. Disciplinel Never discipline your subordinates in front of the other men and women unless you specifically desire to do so for the effect. Otherwise it will just make the one being disciplined very angry with you and his job.

7. Organization and System.

a. You cannot supervise, nor can you eliminate confusion and overlapping of jobs, unless you plan and organize your work. You must have a system to your supervision. You must break down the job, for which you are responsible, into components and evenly spread these job components among your men and women.

b. You, as a supervisor, are responsible for properly supervising your subordinates' training. Each man and woman in your section should not only know his specific duties, but should also be familiar with those of men and women in related sections so that during

SECTION III

RESPONSIBILITIES

1. This section will outline some of your responsibilities as a section supervisor regarding:

a. Section Mission Performance.

b. Supply

c. Ground Safety.

2. Be Familiar With:

a. The mission of your section.

b. Maintenance requirement and who performs this maintenance.

c. Your capabilities and limitations. (What plans for additions or changes in your section's equipment or operation is in progress?)

d. What is needed to perform an even better job.

3. Know Well:

a. Your basic supply requirements.

b. The action to take to obtain supplies.

c. Proper care and use of equipment assigned your section

d. All publications covering your section.

4. You Must:

a. Emphasize preventative maintenance and cleanliness within your section.

b. Make inspections frequently.

c. Require neat, properly prepared, comprehensive records be maintained.

d. Indoctrinate all personnel under your control on the care and safekeeping of government property. This will include security of property to prevent theft or misuse, and the protection of property from the elements to prevent rust or other deterioration.

e. Instruct your men in ground safety procedures related to their **job.**

f. Report every accident immediately to your supervisor and follow-up to insure that it is in turn reported to the Squadron Ground Safety NCO or Officer.

SECTION IV

CONCLUSION

1. Now that you have read this handbook, reread it to make sure that you understand just what you are supposed to do. If you have questions relating to your responsibility, take them up with your supervisor. Keep it handy for reference.

2. Be the best supervisor that you are capable of being by practicing the principles outlined for you in this "Supervisor's Handbook".

Covering Loyal Patriots Backs

In addition we were taught that as a team player loyalty was paramount or as the seals say and as the infantry people say you never leave one of your own behind on the battlefield. As opposed to what happened in Benghazi, a commissioned officer who had gone through OCS would never have allowed those folks to not be at least attempted to be rescued. It is simply unconscionable that Pres. Obama did not try to rescue those folks in Benghazi. The fact that he was trying not to have the United States get involved in another war because of the upcoming election is totally irresponsible and self-serving and not worthy of a president of the United States of America. For a man to put his own self-interest ahead of the interest of the country he is supposed to be leading is traitorous and he should be impeached. This man has no concept of honor, duty or loyalty.

Fannie Mae & Freddie Mac

most of us are well aware of the fact that because of the subprime loans that were issued by Fannie Mae and Freddie Mac and the fact that many of these loans could not be repaid by the people who receive them, that this was the major cause of the economic crisis of 2008 which brought our economy down on his knees. Then when the big Walt Wall Street companies bought up these loans like junk bonds and sold them to other people it just made the situation that much worse. It compounded the problem severely. What happened was many investors wound up with worthless stocks and commodities and lost all their money as result. All this together brought the huge economic downturn or in other words the big bubble broke!

That Obama printed a lot of money into and try to pay off this whole deficit with money the country really didn't have which, of course, with the United States $1 trillion further into debt. Now his answer is to raise taxes on an already destroyed economy saying that he needs to spend more money to get the economy going. It's like somebody who's in a hole trying to dig it deeper in order to get out of it. It makes very little sense logically or economically. But that's the way the man thinks and apparently nothing can be done to change his mind. It makes one wonder if he is really this stupid or is he trying to destroy the United States. Personally I think it's the latter.

International Security and our Defense

Because of his lack of any kind of a military background whatsoever it would appear that Pres. Obama just does not realize that by having a strong nation is the best way to maintain our security in this dangerous world we live in. He does not seem to understand that if we as a nation act strong and tough that we will get much better respect from the rest of the world. We can always be a compassionate nation and try to help those other countries that deserve being helped but we must also totally stand up for human rights and for our own citizens rights throughout the world. We must always appear to the rest of the world as an understanding country but also a tough country not to be fooled with. His total lack of support for the Syrian people while they're being massacred by their president aside is a perfect example of a president not taking action when some action should have been taken. His total seeming lack of conviction when it comes to Iran and their nuclear production also seems to be an indication of his inability to take any action or make any decision with respect to other nations actions. One need only to just consider the Fort Hood massacre where in a major in the Army of the United States who was a medical doctor and a psychiatrist and who was supposed to be counseling returning veterans coming home from the wars overseas instead brutally shot and killed 17 of them and wounded many more in an apparent Islamic jihadist action. This major has never been put on trial or convicted of anything and he still is collecting pay and has already accumulated over $250,000 in his bank account since this all took place several years ago. This shows that either Pres. Obama does not know what's going on or is sympathetic to Islam. Although he purports to be a Christian he certainly does not act like one. His past association with his preacher Jeremiah Wright in his affiliation with Lew Ayres seem to give an indication of his surprising background. The fact that no one can seem to find a copy of his birth certificate, or miss college transcripts, or of any previous girlfriends he may have had as a young man or while in college leave one wondering what his real background consists of.

Money interest rates

When discussing monetary interest rates, the late Nineteenth Century Austrian economist Eugen von Boehm-Bawerk is quoted as saying "The higher are a people's intelligence and moral strength, the lower will be the rate of interest."

If that is indeed true, then our society is in bad shape. At the turn of the century
around 1900 the usury rate in many states was seven percent (7%). In some states if a loan merchant charged above that rate he/she had to pay back the person
borrowing the sum an amount twice that borrowed plus also pay penalties to the
state.

Consumer credit in the United States is most commonly of the type called the
Installment Plan. Installment loans accounted for over a Trillion $ in 2011 and
over 25% of the gross national product. 80 million Americans are now head-over-
heels in debt with installment loans with interest from 14.9% to 29% depending
upon the bank and the credit *card* being used. Bankruptcies are now at epidemic
proportions with no relief in sight. Soon, the amount of available spendable
money will be so low that people will have to stop buying.

Consider the situation in Albania in the late 20th Century. There the government backed pyramid scheme which the majority of Albanian citizens bought into collapsed. The country was under siege from its own citizens for many years. There was wide spread fighting in the streets and virtually all the people were carrying weapons which they were allowed to by the military to steal from the military armories around the country. They were demanding the President resign his office. This is the result of people having no other recourse but violence to affect their desire to survive. Can anyone say that these Albanians are all crazy? I don't think so. Virtually the entire population was involved in this rebellion at the government for having totally misled them.

How long do the banks and loan companies in this country think they can
continue to fleece or bleed the public? When it was suggested by some that the
high interest rates and excessive peddling of credit cards to almost every
breathing human being in the country might be partly responsible for the current
debacle along with the subpar loans backed by FanniMay and FreddieMac that the recent articles in the papers regarding the great increase in bankruptcies and extremely large consumer debt have portrayed, the bankers and our government were indignant. When accused of targeting the most vulnerable consumers, those
with the least capability to pay back these loans, they denied
it vehemently. The facts seem to show differently, of course, with the rate of foreclosures at an all-time high.

What ever happened to the golden rule? "Do unto others as you would have
them do unto you." What ever happened to being kind to people? What ever
happened to being your brother's keeper? Do they all want to live in a world
where it is literally "dog eat dog"? Does that make for a better society? Does that help control crime, reduce drug use?

There was a time when banking institutions encouraged young people to save
money. What a remarkable concept! Now, they discourage people from saving
money. Just to keep a savings account without sustaining "service charges" most

banks require a $300 balance. They give you a less than one percent interest, perhaps 0.6 or 0.7% up to $5000 and then they give you a whopping 1.2%. In the meantime, they are charging you 16% to 30% for credit cards, service charges, and ATM fees, check charges, etc. Soon they may charge you to talk to them in person.

Don't they realize that if they charge exorbitant interest that eventually the spendable money available in the economy is reduced? When people live beyond their means, in other words, when people spend more than they make, it soon catches up with them and something has to give. Yet, still the banks encourage people to use more and more credit cards. That is why the bankruptcies and foreclosures are so high now.

You might think, the banks would reason "well, maybe if we don't charge so
much interest, more people will have happier lives, we'll have more customers,
fewer people will get themselves into trouble and we'll all be happier." But, alas,
that is not the case.

It would seem that mankind is so greedy and self-destructive that eventually we
will destroy ourselves. The following discussion may give this concept more
validity.

Our biosphere

On the one side we have the possible destruction of our biosphere in other words our planet. On the other side we have the radical cause of some religious or radical people who seem to believe that their religion or philosophy is the only true religion or philosophy and that all other religions and philosophies should be done away with. Or put another way if they cannot convert those who they consider nonbelievers then these, the nonbelievers, must be exterminated. This is the basic concept for the Islamic religion. They are not willing to live alongside of other religions. They are not like Christianity or Judaism or Buddhism or any of the other religions who will allow other religions to exist with them. Our Constitution grants us freedom of religion without the interference from the government, it does not allow one religion to take over and do away with all other religions. As far as our Constitution is concerned the Islamists cannot be allowed to have their way in our country. We do not for, example, want sharia law. We do not want to have requirements where women have to be totally covered at all times and are not allowed to talk to anyone other than a close relative unless allowed to by their husband. We do not want to allow women to be stoned in the marketplace because they have been raped or molested in some way. This is not what a civilized law-abiding society allows.

Back to the discussion of our biosphere. We must be careful not to clear-cut all of the rain forests or other areas of the world which produce most of the oxygen which we breathe. We also must be careful not to destroy other life forms to the extent of making them extinct. However we must not go out of our minds pushing this agenda to where we limit the capability to live on this planet comfortably ourselves. In other words we do not want to take protecting the environment to such an extreme that we destroy our populace in our efforts. Many of the so-called tree huggers or greeners are such fanatics that they're willing to kill people to get their own way.

We human beings on this planet have a momentous decision to make which directly affects our very survival as a species. We must decide if we want to live reasonable lives and leave a world capable of living on to our children and their children's children or the alternative. There is only one alternative and that is that we will exterminate ourselves by leaving an uninhabitable world for our posterity if we are not careful. However, being careful does not mean that we should destroy our economy or go to ridiculous extremes. We must try to maintain a balance and not become fanatics.

The extreme in either direction may sound and it should ominous and it should. While these two choices may at first reading seem extreme, when carefully analyzed by using current facts available to us,they emerge as an absolute certainty and will eventually be a fully nonreversible if we go too far in the destruction of our environment.

If you think this is a rather extreme viewpoint, scientists now

conclude
that this destruction of the environment scenario has happened
several times in the past history of our planet. Civilization's
namely destruction has been caused in the past by one or more
of several factors, namely being struck by a meteor or asteroid
from space, lava upheavals causing blackout, disease, or
destruction of the environment and therefore the "Biosphere"
we live on.

   It may take a hundred years or more at the rate we are
propagating the species; destroying our forests; polluting our
drinking water, oceans, lakes, and streams; killing off species of
animals, fish, and foul; exhausting our fossil fuel resources;
creating so much waste product from humans, industry, and
radioactive power processes, and generally destroying our very
own biosphere in which we currently live. Once we reach a
certain specific point, our effort will have become irreversible and
no longer correctable without external help. That, of course, is
virtually not possible. At some point, the environmental systems
on earth will begin to very rapidly accelerate towards total collapse of the
ecosystem as we now know it. Once triggered, the process will happen so quickly
that there will be no hope of stopping it. It will be an out of control, run-away
process. We just need to be careful we don't go that far.

   As far as any help from other worlds or races, to date our
efforts to detect xtraterrestrial life have been fruitless. We need
not expect help from super intelligent beings from beyond our
galaxy. If there were visitors from space, they would of necessity
be very much advanced beyond us. Probably so far advanced
that they would consider us only a nuisance like a swarm of flies,
mosquitoes, or ants and not worth trying to salvage. They might
very well be predators and consider us only prey to destroy.
   If this seems harsh, one need only look back on how more
advanced societies
always seem to treat lesser advanced societies. Remember the
Spaniards and
how they destroyed the Aztecs. The Europeans and how badly
they treated the
American Indians in this country.

   You could reasonably argue that we are well on our way
to reaching the point of no return toward total ecological

collapse. Many species of fish are no longer
present in the fishing areas of the oceans having been
fished to extinction.
Namely flounder, haddock, and others. Many animals are
nearing extinction
throughout the world, namely tigers, elephants, whales,
rhinos, and hippos
among others.

The ozone layer, which used to have a hole in it for only one or
two months
during the year over the Antarctic South Pole now has a hole in it
most of the
year. Already there have been many cases of severe sunburn
caused by
excessive exposure to ultraviolet radiation requiring
hospitalization in the
southern part of South America. This has been a direct result of
strong ultraviolet
rays no longer being blocked out by the ozone in the
atmosphere. One thing which may work to our favor is the fact
that NASA (2013) has now determined that carbon dioxide is
probably not the cause of thermal heating. Rather just the
opposite is true. The carbon dioxide reflects the sun's rays
coming into the atmosphere and helps to keep our planet cooler,
not hotter as originally suggested. This of course upset many
scientists and politicians who had a different agenda there
wishing to pursue. The movie "An Inconvenient Truth" about
global warming, apparently was using faulty data and coming up
with the wrong conclusions. Our ex-vice Pres. is probably not
very happy with that at all.

What should we do? What must we do right now?

First we must decide what actions are causing the worst
effects now. We must prioritize our efforts. We must determine
which actions that are being done right now by man, that are
causing the most harm. Then we must stop as quickly as
possible those actions which are doing the most damage to our
continued existence on this small planet called Earth. We must
take action to do this while evaluating the costs for these actions

versus the results if we do not take them. Even though the costs and consequences may cause us to lose money or status or some of our freedoms, we must very carefully consider what we are going to do to control the situation. It will not be easy, but the alternative is extinction of the species.

Taking the problems one at a time and showing factually the extent of the damage being done or already done and then determining a course of action to stop or reverse the damage would seem to be the most logical way to proceed.

Do you hear any of the politicians or "so called" leaders seriously addressing any of this?

Overpopulation: There are currently over 7.5 billion people on
this planet.
Only three hundred years ago there were approximately one hundred million.
The rate of increase until 1900 was about 1.3% per year. Since then it has
increased to about 7% per year. At this rate the population will be ten billion by
the year 2020. By the year 2040 the number will be one hundred and six billion. Already, even as you read this, many people in many areas of the world are literally starving to death or are severely undernourished and ill. Look at the African countries like Somalia and Rwanda. Look at India, parts of China and other countries. Even in our own country, many millions of children, middle aged, and older people are malnourished. In Mexico, Brazil, Argentina, many children are hungry and also have no place to live. In Brazil and other countries the notorious death squads move around the big cities at night and shoot homeless children and remove them.

What is the solution to the problem of overpopulation? The obvious answer is
take a "time-out" from reproducing or having children for specific period of time until the death rate can overcome the birthrate and maintain that for a determined period of time until we can obtain a decreasing population in the world for another specific period of years to decrease the strain on the

environment and the earth's resources. We need a moratorium on births for a few years. Obviously, no one will want to promote this effort.

Is the above solution possible? Considering the past performance of the
human race one would conclude probably not, but, we could certainly strive to
reduce the numbers as much as possible. Education on the magnitude and
seriousness of the current crisis and methods of birth control world-wide might
help, especially in those areas where the problem appears most acute. Those
areas appear to be China, Africa, India, and some other countries as well. We
are driving ourselves out of room on this our one and only planet home capable
of supporting life as we know it.

As is always the case, the problem is that no amount of writing or words or
yelling and screaming is going to make any difference in how each individual
persons will react to the situation. Most people are so busy working and trying to
make ends meet for their family and so tired at the end of the day that they just
don't have the energy to devote to another "warm and cuddly far-fetched cause".
They just say to themselves "... the extremist are at it again, they must not have
enough to do to keep themselves busy and are dreaming up more world problems and catastrophes."

Virtually all the people will say to themselves, "it's not my problem. Whatever I
do will not amount to a drop in the ocean. I have no control over the situation. I
didn't do it so why should I worry about it. It will not affect my children so much
anyway. Maybe in a few hundred years it will be a problem but

why get excited
about it right now. I have no capability to change the course of human history,
I'm just one little person. What possible impact can I have on the problem, good
or bad?"

And so it goes.

   Until we as a human population on this planet take responsibility for our very
own actions, not much is going to be accomplished to correct and possibly
reverse the ongoing onslaught of the environment.

   A short time ago I heard a man who was discussing the primary river in the
state of Washington on National Public Radio. He was saying how all of the
people and industries along the river agreed that something had to be done or all the salmon would die. But, they could never get together on what to do to stop the process and now there are no more salmon in the river. He was very sadly saying "we knew what was happening and we just let it happen ... we just let it happen, I can't believe it...we watched and just let it happen."

   The very same thing will probably happen to all of us some day in the not too
distant future on this planet. We seem to think that it is somebody else's problem, not ours. Perhaps we forget that there is no place else to go. This is not the 23rd or 24th Century and we are not in a Star Trek Movie, there is no Starfleet Command here. This is it. What you see is what you get, like it or not. We're not going anyplace else. There is no escape. If we screw it up here there is no recovering from it, no remedy. Once we reach the point of no return on fouling up our Ecosystem, we will not be capable of correcting it. We will all die trying. We will become extinct, just like the dinosaurs did 65 million years ago. Perhaps some future race populating the planet a few million years from now will find a few good samples of DNA and try to clone us again and make a movie of that.

   As mentioned earlier, some scientists and others believe that

this complete destroying of all life on the planet has happened periodically throughout the history of this planet many times. Civilizations and animal forms seem to have been completely erased many times before in the planet's history. Some think that the human race is like a parasitic
infestation which will eventually destroy itself to make way for a new group of life
forms which may or may not be more intelligent or beneficial to the planet.
Evidence which has been uncovered seems to support this theory of a cyclic
change in the life forms inhabiting the earth over the last three billion or so years.

Perhaps this is not a bad thing. After all, maybe the situation gets better each
time there is a new occupant. Perhaps eventually some life forms will realize the
value of their environment and resources available and manage to conserve
them and replenish them without destroying themselves first.

Take for example the Rain Forest of Brazil's Amazon Jungle area. It is still
being destroyed at an alarming rate. So far an area twice the size of the state of
California has been destroyed by builders and cattlemen to use for new buildings
and ranging cattle herds. Thousands of species of animals have been made
extinct as a result of the onslaught. Many indigenous natives have been
murdered because they tried to stand up to the trespassers. They were shot or
burned out by these people. Brazilian authorities say they are trying to curb the
exploitation but admit their efforts are not working very well.

The whale population of the oceans is virtually just a vestige of

what it was.
Gone are several species already. Only some blue whales

and humpbacks
remain.

The fishing industry off all shores of the world is virtually

depleted of resources.
Many species are now extinct.

Potable water on the planet has reached its lowest level in our

history. So much fresh water is now polluted with contaminants

that it cannot be economically recovered for human

consumption anymore and the remainder is salt water. New

more economical processes for desalinating water are now

available but they could never keep up with the demand that would

be required if we continue to pollute our water on this planet at

the current rate. New sources

are in dire need of being found.

We have touched on overpopulation. There are too many
births compared to the number of deaths. Another problem is
over clearing of the forests. This is caused by over consumption
of paper products, wood for building materials, and clearing for
additional range or farming land. Computers, which were
supposed to make life simpler for all of us, have quadrupled the
need for paper. Overpopulation has created a much greater
need to build more shelters and offices. Greater population has
created a need for more farm and grazing land and water.

In addition, the increased need for electricity has caused over
use of hydropower causing more damming of rivers and
destruction of natural ecosystems as a result. The overuse of
hydrocarbon products like oil and gasoline has caused greater
pollution of the atmosphere. This overuse of automobiles has
created a vast junk yard of metal around the world. The planet
and its landscape are slowly changing from agriculture to deserts
in some areas. Excessive use of PCB's, CPC's, and other
chemicals have caused poisoning of water sources, destruction
of the protective ozone layer, which now however appears to be
recovering, and many other environmentally damaging

conditions. Over fishing of streams, lakes, and the oceans has caused a severe shortage of fish and fish products as well as the extinction of many species including most whales, and the endangering of many more like the porpoise, dolphin, tuna and salmon.

The accumulation of radioactive waste from atomic reactors has reached alarming proportions with no countries or locations willing to have it disposed of in their territory. You can't blame them, but what to do with it?

You see, man has gotten himself into this because he did not take the time to think it through in the first place. Who in their right mind would use a fuel which is super dangerous and impossible to be disposed of? The only reason it was done in the first place was because it was thought to be an economical and clean way to generate electrical power. No thought or solution was developed at the same time to allow for the proper handling and disposal of the radioactive waste left over.

They get "A" for effort and "F" for forethought.

There are so many powerful vested interests in this country and worldwide out there that this writer seriously doubts that we as a species are going to survive our own greed and deceitfulness. It is sad to say, but the powerful interests are putting instant gratification, greed and the "bottom line" way ahead of reasoned and prudent use of resources and processes.

The only way it can possible be stopped is by us taking positive and strong action. We, of course, will not. The entrenched economic powers are far too great to overcome now.

Who among us has the drive or will or the time and resources to devote to getting the situation corrected? No one! Absolutely no one.

So our economic situation is one side of the story along with the so-called climate change situation. And when you include the self-serving interests of some members of our government as opposed to trying to do what's best for the country they're trying to line their pockets and get reelected we certainly have our work cut out for us.

The above essay is directly related to the initial discussion regarding high and
excessive interest. 'It shows the "mind set" of the average "businessmen". There
is no room for consideration of human needs or the future of humankind when the most important thing in the world is "the bottom line". It may well be humankind's final "bottom line".

Without integrity, honesty and loyalty – what's left?

Shakespeare was certainly correct and prophetic when he wrote, "What fools
these mortals be,"

www.ingramcontent.com/pod-product-compliance
Lightning Source LLC
Chambersburg PA
CBHW071551170526
45166CB00004B/1629